Gus felt hot.

He had a flu bug.

"I cannot run," said Gus.
"I am a sick insect."

Mom put Gus to bed.

"You must drink this broth,"

said Mom.

"And you must rest."

Mom cut up a muffin.
Gus slept.

Dad cut up a napkin.

Gus slept and slept.

At last Gus got up.

"I am not sick!" he said.

But Mom and Dad felt sick.

"You can help us," said Mom.

"This flu bug is bad,"
said Gus.
"You must drink this broth.
And you must rest."

The End